Magnificent
MOROCCO

Dr. Diana Prince

AuthorHouse™
1663 Liberty Drive
Bloomington, IN 47403
www.authorhouse.com
Phone: 833-262-8899

All photos are used with the permission of Getty Photos, except for the Author Photos on Pages 14, 15, 17, 21, 26-29, 37-39, 78-84, 89,109-111, 115-119, 122; 125-127; 129, 130, and 133-143.

This book is printed on acid-free paper.

ISBN: 978-1-6655-4653-9 (sc)
ISBN: 978-1-6655-4654-6 (hc)
ISBN: 978-1-6655-4652-2 (e)

Library of Congress Control Number: 2021924874

Print information available on the last page.

Published by AuthorHouse 01/17/2022

authorHOUSE®

Table of Contents

Introduction

Morocco is a land of incomparable beauty. Located on the northwestern coast of Africa, it casts its spell of adventure from the soaring Atlas Mountains, to the astonishing landscapes of the Sahara Desert.

From the seacoast city of Casablanca, to the Royal cities of Meknes and Rabat, to the bustling medinas of Marrakesh, civilizations have flourished here for centuries.

The sheer beauty of Morocco, and the warmth of its people, make it an unforgettable adventure.

Hassan II Mosque

Casablanca

Casablanca is located on the Atlantic coast of Morocco. Its history of French colonization has resulted in in positive European influence blending with the country's Moorish roots.

At the water's edge is the impressive Hassan II Mosque with its soaring minaret. It is the largest Mosque in the world, able to hold 80,000 people.

The city was first settled in 700 BC for some time by the Romans, and later by the Berber tribes. It's current iteration dates to the 1700's.

Casablanca is bound on the west by the Atlantic, and in the north by the Mediterranean. East of the city lies the desert of Algeria. Southern Morocco is bound by the Western Sahara. The country is an idyllic blend of sea and desert, and it is a region of remarkable beauty.

The desert which comprises a large part of the country is the Sahara. In it's totality, the Sahara covers over three million square miles, and touches eleven African countries.

Casablanca is the largest of all Morocco's cities. In the 12th century, a Berber settlement called Anfa occupied this area. The times were perilous for ships which were often attacked by pirates. The Portuguese drove off the marauders in the late 1400's. They later built a settlement which was known as "Casa Branca", which literally means "white house". It was the origin of the city's name.

The Portuguese settlement endured almost two hundred years, until it was destroyed by a severe earthquake in 1755. The city was not restored until the end of that century by the Moorish adventurer, Sultan Sidi Mohammed.

The newly restored city, with its temperate climate and beauty, brought many Europeans, principally from France and Spain. In the late 1800's, the British were importing wool and tea to Britain.

In 1906 a new French colony was built here. In accordance with the terms of the Treaty of Algeciras, the French were given pre-eminence among the foreign settlers in Morocco. They established *La Compagnie Marocaire* with the Moroccan government. In an exchange for the French funding development in Morocco, they would control incoming customs revenue. Through this investment venture, France agreed to finance the development of Casablanca. Some problems emerged in the French handling of a railway extension. Local citizens rebelled over the situation which caused serious conflicts with the French. The situation devolved as troops were stationed within the town, and led to conflict with locals. There were some casualties in the conflict.

However, by 1912, conditions improved when the French Protectorate was established at Casablanca, and it endured for about three decades. The negotiations were laid out in the Treaty of Fes.

Newcomers from France arrived in great numbers, and for a time, they comprised over forty percent of the population in Casablanca.

During World War II in 1943, representatives from the United States and Britain met at the city for the Casablanca Conference. Allied leaders Franklin Roosevelt, General Charles de Gaulle, Winston Churchill, and General Henri Giraud met in Casablanca in January of 1943.

The existing French Protectorate of the French in Casablanca officially ended in 1956, when Morocco officially became independent from France.

Today the city is still popular. The stunning coastline is near Boulevard Hansali, which is now a famous tourist shopping area. The old city Medina lies further inland. The rustic ramparts enclose the oldest part of the city, which is a tourist mecca. Here original stone walls and narrow streets of the old city still have a timeless grace.

The district around the Muhammed V Square, located in the old city, was built primarily by the French. From this square, avenues reach in all directions from the vibrant, and usually crowded center. The outlying areas around the square extend into the areas of business and commerce, with large shops, hotels and parks. In the distance is the Catholic Sacre Coeur cathedral.

There are two major airports--the Anfa Airport and the Nouaceur Airport, as well as a railway line to Tangier. The Mohammed V International Airport is a modern terminal. The main office of the Royal Air Maroc Airline is located at the Anfa Airport.

The Straits of Gibralter separate Morocco and Spain. At this waterway, this water corridor has only a nine-mile distance between Europe and Africa.

The El Hank Lighthouse was built at Casablanca in 1916. It is located at the Boulevard de la Corniche. It was designed by Albert Laprade, a French architect. The lighthouse is the highest in all of Morocco. The light has a visibility range of 30 nautical miles. There is an excellent view from the top of the 167-foot tower. Besides the sweeping view of the coast, the lighthouse provides an excellent view of the Hassan II mosque and its famous minaret.

The lighthouse has 256 stairs. Visitors can climb the spiral staircase to the viewing deck at the top.

Another popular attraction for visitors is the Sacre Coeur Cathedral located at Rue d'Alger and Rachdi Boulevard. Designed by architect Paul Tournon, it combined a new-Gothic style architecture with some art deco design elements. It even incorporated some small architectural touches of Moroccan design which blended surprisingly well with the more formal architecture. The Cathedral was opened in 1930.

After Morocco's independence from France in 1956, the Sacre Coeur Cathedral became a multi-purpose building for the staging of exhibitions and other cultural events.

One enduring visitor favorite in this city is the famous "Rick's Café". The 1942 movie classic, "Casablanca", had featured Ingrid Bergman and Humphrey Bogart, and it became one of the most beloved films of all time. Despite the fact that the film was completely filmed in a Hollywood studio, it became the incentive for building this restaurant which has become a tourist favorite. Inside, the owners have replicated the interior of the film's set, and brought the movie's romance and nostalgia to thousands of visitors who have made it a "must see" landmark on their visit to Morocco.

The Sindibad Theme Park Casablanca is an unexpected sight in Casablanca. It is an amusement park with rides including a roller coaster, children whirling in giant teacups, and even small trains for children to ride. There is also a zoo with real tigers, giraffes, elephants and lions. There is even a lush green space on the expansive grounds for family picnics.

Most Moroccans in Casablanca live in the extended city neighborhoods, some live in small outlying settlements in the nearby desert.

Casablanca is the economic and commercial center of the country. The Royal Moroccan Navy is headquartered in Casablanca. Today, over three million people live within the city of Casablanca, and an additional four million live in the expanded, outlying metropolitan area. It is the largest of Morocco's cities. The total population of the country of Morocco is 38 million. Twenty million of these are Berbers.

The two predominant languages of Casablanca are Arabic and Tamazight.

The current ruling monarch in Morocco is Mohammed VI.

Couple on the Shore at Casablanca

Sunset at Hassan II Mosque

Hassan II Mosque is one of Morocco's largest.

Interior of Hassan II Mosque

Mosque at Sunrise

Fishing Boats at the Pier

Fishermen unload their catch at sunset at Port de Peche.

Morning Walk in Casablanca

Rick's Café at Casablanca

Stock Exchange in`Casablanca

Afternoon Tea

Moroccan Man in Ouarzazate

Ouarzazate

Ouarzazate lies south of the Atlas Mountains at an altitude of 3,800 feet. The city of 70,000 is a launching point for expeditions into the vast Sahara Desert. It is located in the Draa-Tafilalet region of Morocco.

The Berbers are the largest population group in this city known as the "Door to the Sahara". A landmark in this region is the village of Ait Ben Haddou on the outskirts of the city. It is known for its impressive fortified buildings with their deep red-colored walls, made from the local red clay in this region. Ait Ben Haddou is on the main route to Marrakesh.

Many excursions into the Sahara leave from this small village on the outskirts of Ait Ben Haddou. This was once the main launching station for camel caravans to Timbuktu in what is now Algeria. It was also the point of departure to the remote desert towns of Merzouga and Tinghir.

Ouarzazate was a historic stop along the ancient caravan routes. The film "Lawrence of Arabia" was filmed here in 1962. Other desert films are still produced here in the town's two large film studios--Atlas Film Studios and the CLA Film Studio. Other movies filmed in this region include "Jewel of the Nile" and "The Man Who Would Be King".

The town was an intermediate point for traders heading to Europe. Sheikh Abu Ahmed Abdullah controlled this region in the sixteenth century.

The huge and imposing Taourirt Kasbah was built as a palace in the nineteenth century.

In Ouarzazate, there is also the Gazelle Animal Natural Reserve of Bouljir. It is dedicated to saving the endangered Dorcas gazelle population. The Reserve has been studying the severely reduced population of the animal. Genetic studies are being conducted to determine ways to reverse the trend.

The Iguerman Nature Reserve and the Tizgui Waterfall are also popular tourist sites.

Ancient Moroccan city of Ait Benhaddou

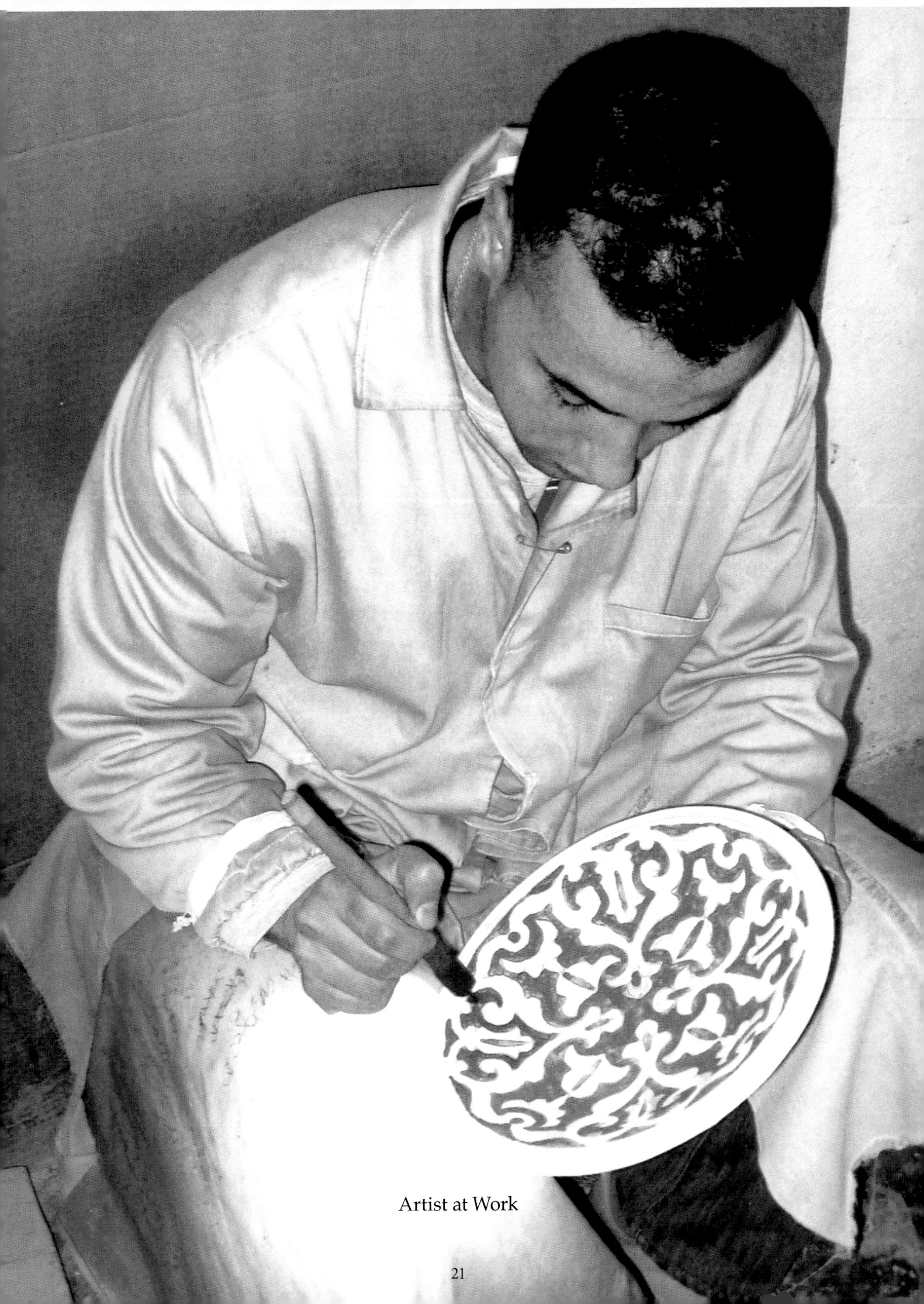

Artist at Work

Casbah at Ait Benhaddou

Camel Caravan in the Sahara

Making Tea at a Desert Campfire at Sunrise

This woman, in a Berber kitchen at Ait Benhaddou,
is cooking flat bread in her earthen oven.

View from Desert Tent

Woman in Desert Tent

Resting Camels

Sahara Dunes

The coast at Tangier

Tangier

Tangier, located at the entrance to the Strait of Gibraltar, dates back to the Phoenicians who arrived there in the tenth century, and were among the earliest to settle here. Many other cultures were to flourish here.

The current name of the city is derived from the Carthaginian word "Tenga". The Berbers, who followed, called it "Tingi", and the Greeks called it "Tingis".

After the Phoenicians left, the Carthaginian traders arrived in the fifteenth century.

The Greeks also established settlements here. Their myths claim that the legendary Hercules came to what is now Cape Spartel, on the coast of Tangier, when he performed his historic deeds of superhuman strength. Today, visitors can visit the "Caves of Hercules", where the legendary hero once rested. These caves are located on the coast at Cape Spartel.

Later, a Roman colony existed at the present site of Tangier. Christians were persecuted around 300 AD under the command of the Roman Emperor Diocletian. Here Saint Casian and Saint Marcellus were beheaded for their faith.

In the fourth century, Spanish armies invaded North Africa. They were later driven out by local Berber tribes. Nevertheless, because of this port's ideal location as a gateway to Africa, factions battled for three hundred years, sometimes with the Europeans who demanded that taxes be paid, sometimes not with money, but with female slaves.

The Almoravids in the tenth century, briefly held power at Tangier. Incursions into this territory continued from many regions, resulting in a continuing instability. The inhabitants were assaulted with one war after another. Eventually lawlessness became the rule and the city was rife with crime and piracy. At one point, Tangier was blockaded by the British for the rampant piracy.

What was never in dispute was that this city was strategic and valuable to all Europeans. As the "Door to Africa", due to its location near Gibralter, it was a vital port.

In 1912, Morocco was divided into Spanish Morocco in the north, and French Morocco in the interior of the country.

In 1923 there was a common zone at Tangier designated as "international" and managed by Britain, France and Spain. The terms of this agreement were ratified by all three countries, who were called the "three partner nations." This international agreement led, for some time, to Tangier becoming a city in which Jews, Muslims and Christians lived together in what was called the "International Zone of Tangier".

After World War II, the residents were still living in the designated "International Zone of Tangier". When that zone was abolished in July of 1952, public records indicated that the population then numbered 40,000 Muslims, 31,000 Christians and 15,000 Jews. These public records also confirmed that despite the great disparity of religious beliefs, the citizens had co-existed very peaceably and with great tolerance for many years.

In 1956, the country of Morocco was also restored to its singular sovereign state as the French Protectorate ended.

Today, in Tangier, nearly one million people have made this city their home.

The former Dar el Makhzen, the sultan's palace, has been converted to a museum. The Moroccan royal family retreats to Tangier during the summer, where there is a royal residence that they use as a second home.

Today, the two official languages are Berber and Arabic. French is also widely spoken in this city. Spanish is also spoken because of the city's long and historic ties with nearby Spain.

Today, the city is a major tourist center with its old medina near trendy modern shops. The city also has a rail system with Rabat, Casablanca and Fes.

Riding Horses on the beach at Tangier

Cape Spartel Lighthouse overlooking the entrance to the Strait of Gibraltar, separating the Atlantic Ocean and the Mediterranean Sea.

Sunset at the Marina on Tangier Island

The Caves of Hercules are natural calcareous caves at Cape Spartel. Legend says that the Roman God, Hercules, once slept here. This occurred, according to legend, when he went on a sacred journey to find the famous golden apples in the Hesperides Garden.

Tangier Marketplace

Young village girl in Tangier

Young Boys Playing in Village

Gazbo Island

Rabat

Rabat, Morocco's capital city, is located in the northwestern part of the country. It is a port on the Atlantic Ocean, located at the mouth of the Bou Regreg River. A small town named Sale lies on the opposite shore nearby. The name "Rabat" means "fortress" or "stronghold" in the native language.

The city's mild Mediterranean climate and idyllic seascapes make it an ideal tourist Mecca year round. The town was originally a military outpost founded in the twelfth century by the Almohads. It flourished for the next five centuries until the decline of the Almohads, when it was overtaken by Barbary pirates who seized ships and caused widespread destruction.

Pirate ships attacked with impunity, threatening commerce and the entire shipping industry. No ships were safe from their attacks and looting.

It was not until much later in the early 1900's when the French came to Morocco, and established a protectorate, that the region thrived. At that time the French established a major naval base there.

In 1955, Morocco was restored to self-government. At that time, Mohammed V moved the capital from Fes to the city of Rabat.

In the 1960's, the Royal Moroccan Air Force was created at Rabat.

The most important city landmark is the Hassan Tower dating to the twelfth century. Below the tower, lie the remains of the ancient mosque originally built by Moulay Yacoub. It was designed to be the largest mosque in the world. However, when he died the construction also stopped, and the massive project fell into disrepair.

Today, the main pillars cover a broad open plaza emphasizing the immense size that had been planned for the massive mosque. Today, tourists wander through the abandoned ramparts. Mohammed V, who was never to see his dream of the great mosque, is buried with his two sons in front of the unfinished minaret. Today, the Mausoleum of Mohammed V is a sacred spot for pilgrims.

There are five main gates to the city. The most famous of these is the Bab El-Had Gate, which is also called the Sunday Gate. The name "Bab El-Had" means literally "the Edge of the Sword."

The charm of the Old Town includes the narrow winding streets of the medina with the seaside colors of walls painted white and ocean blue.

The residence of King Hassan II and the royal family is the "King's Palace". The palace is named "El Mechouas Essaid", which means "the way of happiness." Many of the government offices are incorporated into the vast complex.

Rabat has a diverse culture of many languages. The basic languages are the generic Arabic and the Darija, or Tamazig, languages spoken locally. Additionally, the native Berber languages are also common in the country. French which is also used, particularly in the universities, dates to the long involvement, history and colonization by the French.

Architecture and modern enclaves in the city are some of the most impressive among all the African countries. Hoy Riad is one of the more prosperous suburbs set apart from the middle class neighborhoods in Rabat, with its large and spacious homes. Other neighborhoods are also well kept in the diverse city.

The city's greatest museum is housed in the former Royal Palace.

Situated on a high cliff is the Kasbah of the Udayas. It marks the place where the city was originally built.

At the Kasbah des Oudaias, lie the lush Andalusian Gardens. They were originally built by the French in the 1900's. Nearby, there is a scenic lighthouse overlooking the large cemetery and the blue Atlantic.

At the ruins of the old Royal Fort are the remains of elaborate and extensive gardens which date to the time of the early French settlers.

The Chella Mineret is located at the River on the outskirts of Rabat. The Bou Regreg River flows nearby. Here time has also left its mark. Ten centuries ago, this small Roman settlement of Chella was abandoned. Storks now inhabit the ruins.

There is a live theatre venue called the Theater Mohammed V which draws large crowds into its center city location. Local tourists also enjoy the Rabat Archeological Museum nearby.

For two decades tourists have enjoyed "The Apartment"--a gallery for artists, which showcases local talent, and also the work of great artists who exhibit their work here.

Another signature event in Rabat is the "Mawazine"--an international music festival which features major worldwide performers. This has welcomed over three million music lovers. This venue not only exhibits visual arts, but also features dancers and other specialty performers.

Despite its heavily Muslim culture and magnificent mosques, there are also Christian Churches established by Protestant evangelicals. There are also Catholic churches in the city. The largest is the Saint-Pierre Cathedral in Rabat. The city is also the home of the synagogue of Rabbi Shalom Zawi.

Sports are popular in Rabat. The city has six primary football teams including such competitors as the ASFAR football club, the Stade Marocain, and the team of Hilal de Rabat. A modern sports venue, built in 1983, is the Prince Moulay Abdellah Stadium. The venue seats over 50,000 spectators. Other popular competitive sports include professional volleyball and basketball teams.

Rabat has an efficient city train system. From Rabat's central rail station, trains head north to Tangier, and south to Marrakesh and Casablanca. The eastern rail line connects with the cities of Fes and Meknes in the east.

This is also a city for walkers. Here and there, street musicians play a guitar-like instrument called the *Gnawa.*

Among the cultural highlights are small cozy restaurants with such Moroccan specialties as lamb Tajne, roasted almonds and chicken pastille.

Storks living in the Chellah--the walled city in Rabat

Rabat Coastline

Lighthouse at Rabat on the Atlantic

Sale Beach is located just north of Rabat.

Women perform their *Fantasia* equestrian horse skills at the Royal Polo Club "Dar Es-Salam" in Rabat.

Suspension Bridge in Rabat

The Medina in Rabat

Boulevard in Rabat

Woman in traditional dress at a folklore festival

Marrakesh

Marrakesh is the quintessential experience of Morocco. It was one of the four imperial cities of Morocco. Like the others, it has regal palaces and spacious gardens. From inside the mosques the haunting chants fill the air. It has some of the most appealing souks with colorful and unique arts and crafts.

The origin and meaning of the word "Marrakesh" is contested. However, in the Amazigh etymology, the Berber word for Marrakesh has been interpreted as the "Land of God." This is based on the Berber derivative word "Mrrakc".

Also, in Medieval times, the entirety of all of Morocco was designated in maps and narratives as the "Kingdom of Marrakesh", and appeared on the maps as "Murakush".

With almost a million people, it is Morocco's fourth largest city. The foothills of the Atlas mountains rise to the east of the city. Berber tribes have lived here for over a thousand years. Marrakesh became an imperial city of Morocco in 1070, as the center of the Almoravid Empire. It is sometimes called "The Red City".

A renewal occurred in Marrakesh in the sixteenth century when the Saadian sultans built elaborate monuments such as the El Badi Palace.

Beginning in 1912, all of Morocco was a protectorate of France. That era ended in 1956 when the Moroccan sovereignty was restored.

Marrakesh is known for its fine universities. It also has impressive architecture like Bab Agnaou, the Bahia Palace and the El Badi Palace.

The largest public square, not only in Morocco, but in all of Africa, is the Jemaa el Fna. It is not so much a place, as it is an experience. In the daytime it is alive with activity. It is a spacious chaotic blend of people. While purchasing treasures, the visitor can stop and watch as a snake charmer coaches a cobra to respond to the quivering notes of an ancient musical instrument. As the clear notes fill the air, the snake moves noiselessly in the still air in response.

The exotic square of Jemaa el Fna never sleeps. As dusk descends, the square transforms as people gravitate toward the activity of merchants, fortunetellers, performers, and shoppers. As night falls, the square is alive with the spicy odors of exotic foods, singers and holy men. Smells of local open-air cooking draw the evening crowds, as smoke from the vendors wafts upward in pale clouds. There is no place like Marrakesh after dark. It is magical.

Above the city is the slender minaret that towers over the Koutoubia Mosque, which dates to the twelfth century. It is a major landmark in the city.

The city was founded almost one thousand years ago, and it retains its original appeal of the exotic and the unique. The souks, which are the local markets, have everything conceivable. Leather products, ornate jewelry, glassware, and unique art are just a short list of what is available in these colorful medinas. For a moment it is hard to take in the full impact of the place. It is alive with activity and color, and a rare energy that is perceptible.

In the 1100's, an elaborate underground water system was built. It was an innovation that was remarkable for the time. In addition to its efficient water system for use by the people, it also provided water for the elaborate and expansive gardens of the King's palace.

The Almoravids were later displaced by the Almohads, an Islamic tribe from the Atlas Mountains. This occurred during the middle of the twelfth century at a great cost of life. Almost 7,000 people perished in the conflict.

The Almohads were orthodox Islamists. They built the Koutoubia Mosque, which is today a symbol and focal point of Marrakesh.

During the thirteenth century, the Almohad tribes began to compete with one another. For over a decade, the various Almohad groups fought among themselves resulting in many casualties.

Ultimately, as the Almohad tribes fought internally, another tribe took advantage of the ongoing conflicts. The tribe called "Zenata" claimed an easy victory as the Almohads fought among themselves.

It was three hundred years before Marrakesh was to rise again, under the Saadian sultans. With the Saadian rulers, some of the majesty of Marrakesh was restored. The new rulers built the elaborate and beautiful El Badi Palace in the late 1500's.

In the 1900's, political revolts derailed the government, and tribal rivalries resurfaced. Amid this chaos, the French arrived. They declared a French Protectorate in 1912, which lasted until March of 1956. At that time, self governance was restored to Morocco.

From that troubled and tumultuous time, Marrakesh, and all of Morocco, emerged into a time of relative peace.

The current ruler, Mohammed VI, reigns over a peaceful nation, and Marrakesh is flourishing as an attractive destination for tourists.

Perhaps one of the best ways to describe Marrakesh is "the unexpected", meant in the most positive way. The local people are respectful and genuinely welcoming to tourists. Generally speaking, it is less expensive than other Moroccan cities. Spring is the ideal time to visit Marrakesh. "Must sees" include the Mamounia and its elegant gardens. A great local culinary choice is the Namaskar Palace.

The Jardin Majorelle, is an unusual but elegant estate adjacent to the Yves Saint Laurent Mansion. The Jardin Majorelle was originally built by an architect named Jacques Majorelle. When it fell into disrepair, Yves Saint Laurent, a world-renowned French designer, restored the elegant estate.

Yves Saint Laurent purchased the estate, and transformed it with bold and striking rich blue accents, and created a work of architectural art. He also built an art gallery featuring Islamic art. He renovated the gardens with touches of that same intense blue signature color for a provocative, but beautiful effect. For almost seventy years, visitors have come to see the elaborate grounds of the Majorelle Garden.

The Mellah, built in the 1500's, is the Jewish Quarter of Marrakesh. Built under the rule of the Saadian kings, it flourished in the sixteenth century with its distinctive synagogues. Today, it is also the location for the Lazama Synagogue, adjacent to a Jewish cemetery.

Man at a Forge in the Medina

Hand in hand at Marrakesh Market

El Badi Palace at Marrakesh

Young boys play in medina in Marrakesh

Spice Market in Marrakesh

Majorelle Garden in Marrakesh

Bab Agnaou at dusk in Marrakesh

Outdoor Rug Market at Marrakesh

Tent in Moroccan desert near Marrakesh

Camels near Marrakesh

Snake Charmer at Jemaa el Fna Square

Dusk falls on Jemaa el Fna Square, Africa's largest square.

People fill Jemaa El Fna Square at night.

Celebrating at Jemaa el Fna Square at Night

Dyer's Souk at Marrakesh

Ouzoud Waterfall near Marrakesh

Fes ramparts against a stormy sky

Fes

The city of Fes in northern Morocco lies east of the Atlas mountains. It is the country's second largest city. It is one of Morocco's four "Imperial Cities".

Located on the old Sahara trade route, for centuries it has been a crossroad for travelers from Marrakesh, Casablanca and Tangier. The city is built on the Fes River, and it is surrounded by hills. It was originally settled in the ninth century.

Later the Almoravids arrived, and under their control the city flourished. In the 1400's, the Marinid sultan, Abu Yaqub, built magnificent buildings, including the royal residence. A distinctive architectural style emerged. Also a Jewish quarter called the Mellah flourished in the city. When the Marinid dynasty was defeated, the city declined as a political power.

Fes still maintains the University of Al Quaraouiyine. It is the world's oldest and still operating university. It was built in the ninth century.

The oldest leather tannery in the world is located in Fes. It is the Chouara Tannery which was originally founded 900 years ago. The traditional tanners use large round stone vats for the brilliant and endless dyes in rich colors. Today three major tanneries exist to produce dyed leather products of the highest quality, which are exported worldwide.

Today the Oued Bou River, which runs through the city, continues to provide a source of clean water. This has been a major asset for sustaining the city and its industries.

Adventurers and newcomers have been drawn to this region since the earliest time. Through the ages Berbers and Arab warriors built homes here. Small Jewish and even Christian enclaves were able to exist side by side.

While the original tribes shared the city in relative peace, the various family tribes, which later emerged to claim the region, often resorted to war, turmoil and even bloodshed to retain power.

From the middle of the tenth century, until the end of the fifteenth century, the main tribal families each sought supremacy in the politics and power of Fes. The Almoravids, the Almohads and the Marinids each flourished in succession.

In the sixteenth century, the Saadi Sultan claimed Fes. He fended off a later attack by the Ottomans. But ultimately, a later coalition of local tribal families sided with the Ottomans against the Saadi rulers, and they were driven out of power.

As many factions reached for power, the ongoing wars and conflicts persisted. Many tribes challenged other factions of their own family for power. After contending with foreigners, the Udayas and the Alaouites challenged each other.

By the end of the 1700's, as the sultans lost the confidence of the people, the Alaouites offered relief with their promise of building and restoring the city. One of their projects was to restore the Royal Palace and its grounds.

From 1912, the French held colonial rule, which lasted until 1956 when Morocco declared its independence.

The two main sections of the city reflect two distinct parts of Moroccan history. The *Fes-el-Bali* refers to the oldest part of the city immediately adjacent to the river. This part of Fes dates back to the ninth century.

The *Ville Nouvelle* refers to the newer district, built years later on higher ground by the French.

Singers perform at the Festival of Sufi Culture in Fes, Morocco.

Young children in a classroom inside the Medina at Fes

The Blue Gate

Young Girl in Fes

Large dye vats are used for making leather products.

Young Boys enjoying games

Woman baking bread in her kitchen

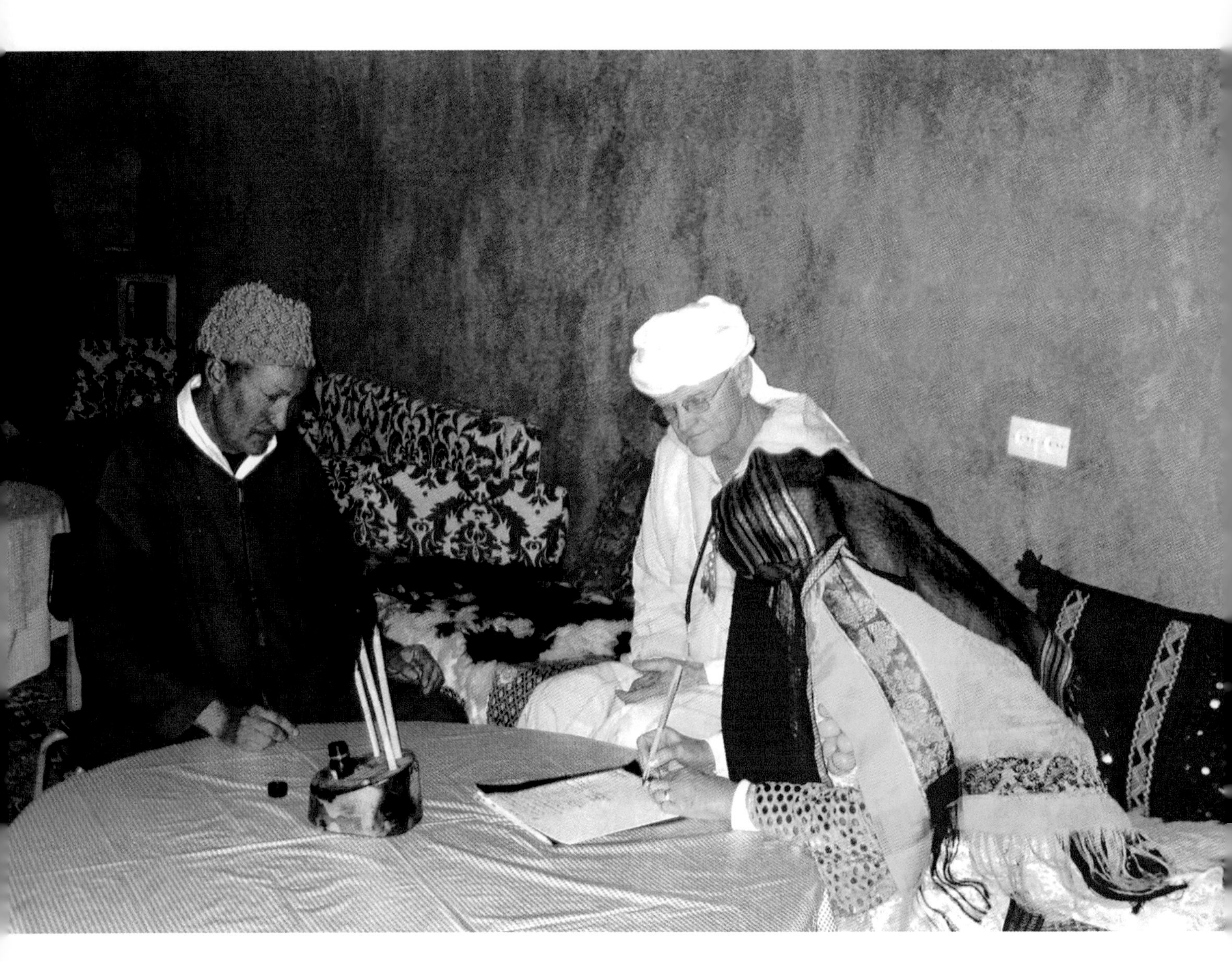

Signing marriage papers

Preparing afternoon tea

Member of the Royal Guard at Meknes

Meknes

Meknes is located near the ruins of the ancient Roman city of Volubilis, which dates back over a thousand years.

The name of the city of Meknes was derived from a Berber tribe named "Miknasa." They are believed to have migrated to this region from Tunisia in the eighth century.

In the 12th century, the Amoravids built the Grand Mosque in Meknes. Inside this former imperial city lies the Bab Mansour gate, with its intricate and ornate arches.

The city of Meknes was originally established by the Almohads in the eleventh century. It began as a military settlement.

In 1672 this city became the Moroccan capital under the Sultan Moulay Ismail. He had been born at Rissani, Morocco in 1645. He was to become the longest ruling leader in the history of Morocco, serving as sultan for over half a century. He was only twenty-six years old when he ascended the throne.

Moulay was second in line to the Alawite dynasty, and declared himself to be a direct descendant of Mohammed. He built the new imperial palace with its massive gates and fortifications. It was a structural wonder in its time. During his half-century as sultan, he built up his lavish royal palace. He also built a large stable at Meknes to house his 12,000 horses. His large-scale architectural projects were often built with the use of captured Christians as slaves.

Sultan Moulay's first order of business, upon becoming ruler, was to move the capital of Morocco from Fes to Meknes. He used Muslim slaves to do this. The council of Islamic scholars called the "Ulema" condemned his use of these slaves as an affront to his own religion. One reason he chose Meknes as the new capital was to escape the censure and complaints from other Muslim leaders in Fes.

Today, he lies entombed inside the Mausoleum that bears his name at Meknes. It is an impressive structure with an ornate outer courtyard, and an interior with fine marble.

During his long reign, however, Moulay became a figure of intrigue, and was noted for his use of terror in his regime. He terrorized not only his enemies, but also some of his own people. This man, who created the splendor of the new Imperial Palace, was feared for his reputation. He is believed to have executed nearly 25,000 people, often for no reason.

Moulay had four wives and 500 concubines. According to the *Guiness Book of World Records*, he was reported to have fathered over 800 sons--more than any other individual in history. Later estimates suggested that the number was over 1,100 sons. This did not include any female babies. If the babies born were female, they were strangled at birth.

When Moulay was warned that one of his favorite sons was planning to seize the throne, the sultan killed him.

He was reputed to have had a dungeon prison below the city of Meknes. The prisoners were mostly Christians captured from ships by Barbary pirates, and they were used as slaves to build the new city.

Upon Moulay's death, his sons fought among themselves for power. In the mid 1700's, an earthquake caused severe damage to the imperial city, and left devastation. The country was to suffer long-term instability. For several years, tribal rivalries and warring factions followed.

Real progress and rebuilding did not occur for several years at Meknes. It was not until the French began their protectorate of Morocco in 1912, that progress was made to restore Meknes. Much of the construction, however, was done north of the old city. Today near the former royal palace, there is a large public gathering place called El-Hedim Square.

Old Granery at Meknes

Fantasia horse riders perform at Meknes. Their traditional horsemanship skills are performed at festivals and Berber weddings.

Women Entering the Medina

Port of Essaouira

Essaouira

Picturesque Essaouira is a port on the Atlantic. The region has been occupied since prehistoric times. The Portuguese settled here in the early 1500's. The Old Medina is the Skala de la Kasbah built by the Europeans in the 1700's.

Essaouira is also a resort town for water sports such as board surfing and wind surfing.

A majority of the people are Berbers. The meaning of the name "Essaouira" is "fortress" or "ramparts". Until this past century, the city was called Mogador--the name originally used by the Portuguese. The island which lies just off the coast is still known as Mogador.

During the third century, the Romans established a settlement on the offshore island. Several artifacts from that colony are currently displayed in the city's local museum.

However, there is evidence that a large settlement existed as early as the eleventh century, and engaged in trade with foreigners.

In the 1500's, other incursions from Europe tried to intrude on the lucrative trade of sugar. However, other nations were kept at bay by pirates who disrupted plans of other nations who wanted to settle here and conduct business.

The city is known for a rare dye found in local purpura shells. It was exported to ancient Rome to decorate the imperial cloaks worn by royalty.

In the early 1600's, the French attempted to establish a base at Essaouira. Admiral de Razilly brought a fleet of seven French ships into the port, and landed at the Mogador fortress. They were turned back. However, the following year, they returned with a more conciliatory approach, and signed a treaty with the ruler Abd el Malek II. They were able to build a mutually agreed upon trade agreement, which benefitted both France and Morocco.

Today the city has a population of 50,000. The Essaouira citadel at the coast rises high above the Scala harbor.

The town was renovated in 1760 by Mohammed III. Well aware of competition from other nearby trading sites in Agadir and Marrakesh, the king hired French engineers and architects to build a newer and more modern city.

This venture succeeded splendidly for well over one hundred years. The port was the most important and prosperous in all of Morocco. Camel caravans, which were routed through the Sahara desert, and the intervening towns of Timbuktu and Marrakesh, made this port one of the most prosperous in all of Africa.

The Moroccan king welcomed Jews to handle the commerce in the city. In the Jewish quarter, many synagogues were built. The trade, however, later decreased when alternate shipping routes in other ports were in competition with Essaouira and the former caravan routes which supplied it. Also, when the state of Israel was founded, many Jews left for new lives in Israel.

In the early 1800's, Essaouira surpassed Rabat with its great volume of exports. Also, at that time, exports from the caravans in Marrakesh were being re-routed through Essaouira.

When Morocco was allied with Algeria against France, the town of Essaouira was bombed by the French in 1844. However, by 1912 Essaouira was made part of the French protectorate.

Today, years later, the French-Moroccan school that existed in Essaouira can still be seen. Also, the French language is still commonly spoken as a second language in this city.

Ramparts of Skala de la Ville at Essaouira

The Beach at Essaouira

Coastline view of Essaouira

Young Boy at the Sea Coast

A Day at the Beach

Seagulls on the Coast

The Gnaoua World Music Festival at Essaouira is held each year. It brings local and international musicians to the city.

Surfing in the north Atlantic Ocean at Agadir

Agadir

Agadir, a coastal city, is popular with tourists. Wide walkways along the coast have splendid water views. The pristine beaches attract sports enthusiasts for surfing and other activities. Cozy restaurants and interesting shops cater to tourists.

The main language here is Berber. The old Kasbah with its graceful arches is the city's oldest building. The city lies near the foothills of the Atlas Mountains. The climate is mild year round.

Corniche La Cote is the most popular beach, attracting surfers and sunbathers. Including the metropolitan area, the city has a population of nearly one million people.

The name of the city came from the Phoenicians who built the city around 1000 BC. "Agadir" means "walled fortress".

Berber tribes occupied the region in the twelfth century, and it was a significant port by the fourteenth century. The Portuguese arrived and built a fort called Santa Cruz in 1505. However, in 1541, local tribes blockaded the fort, and later captured six hundred Portuguese living there. Eventually, the Portuguese abandoned this region, except for their previous settlement in Tangier.

The Casbah was built in 1572. By the seventeenth century, the Tazerwalt Berber dynasty had built a significant trading harbor, primarily in copper and sugar exports.

In 1731 a severe earthquake destroyed the city, and trading shifted to what later became Essaouira. In the next three decades, Agadir was eclipsed by Essaouira.

The decline in commerce lasted more than a century when Agadir's population dwindled and the city's buildings fell into serious disrepair. It was not until Agadir was made a French protectorate in 1911, that the city began to rebuild, and ultimately to thrive.

Beginning in 1930, the French rebuilt the seaport and the city's infrastructure. Fishing and agriculture flourished. Mining operations began in the region.

In 1960 disaster struck. An earthquake killed almost forty percent of the city's population. More than 16,000 people perished in the disaster.

A massive rebuilding effort began. Reconstruction moved the city one mile south of its former location. Today it is thriving. It is also the most important fishing port in Morocco.

Additionally, mining for manganese and cobalt have provided a lucrative industry in this city, and helped to revitalize the city.

The El Had souk is the largest in the city with over five thousand thriving shops.

Today there is an excellent view of the sea and the city from the Casbah called Agadir Ou Pella. Visitors also flock to beautiful Legzira Beach with its stunning red natural arches, a visitor destination located south of Agadir.

In the eastern part of the city, there is a large botanical park with rare plant specimens. Adjacent to that park is the Agadir Crocodile Reserve. There visitors can observe the rare Nile crocodile species, in its natural habitat.

On the beach at Taghazoat in Agadir, Morocco

Old and new on Taghazout Beach in Morocco

Arch located south of Agadir

Riding horses on the beach at Agadir

Closeup of Ifri Granary rooms in Taliquine province in Southern Morocco

Granary rooms and chambers at Agadir

Passageways in the ancient city of Agadir

Children play in the oldest part of the city.

Two boys play among the ancient buildings of Agadir.

Climbing the Todra Gorge near Dades in the Atlas Mountains

Todra and Dades Gorges

The Dades Gorge is located about thirty-two miles from Tinegar. It is also about seventy miles northeast of Ouarzazate. It is a stunning steep rock gorge carved by the Dades River. The river source lies in the highest ranges of the Atlas Mountains. Over the centuries that river source has gradually created some of the steepest rock terrain on earth.

The Dades Valley with its steep mountain walls has an amazing diversity. At certain times of year the desert can stretch in one direction, while snow is falling in the opposite direction.

Even more surprising is the fact that engineers built a narrow, winding one-lane highway clinging to the steep mountain curves. It is considered "the most dangerous road in the world." The highway clings to the steep mountains like a winding ribbon. The hairpin curves attract daredevils from all over the world.

This writer first heard the story as I sat securely in an old bus with about twelve other travelers. We had been rattling along the dusty roads since leaving Tinegar that morning. The driver announced that the gorge walls were almost 1,700 feet in height. These gorges were formed millions of years ago when tectonic forces created the lofty Atlas Mountains. At that time this land was still under the ancient ocean.

The road on which we were driving is designated on maps as "Route 704", and is also known as the "Road of a Thousand Casbahs." The driver then announced that the clearance of our vehicle was only 12 inches on the side of the road which had a steep drop down thousands of feet. The passengers gave a collective gasp. We had already started up the steep grade to the heights when he finished.

Suddenly, we saw the road seem to fall away on one side as we peered straight down into a chasm below us. The hairpin turns needed to be navigated with surgical precision. We could see that the road was astonishingly narrow, with the steep plunge to the right, and a clearance of only one foot to the edge.

Suddenly, there was a stunned silence in the bus, where casual conversations had been going on a few minutes earlier. In the window seat I looked out, and I could not even see the 12-inch clearance between us and a free-fall. I saw nothing but an endless drop of thousands of feet.

I then experienced a moment, not of fear, but of abject terror. Everyone went deadly silent.

"What if someone comes around the corner in the other direction?" one man called out.

"Well," the driver said, "We *hope* that won't happen, because there is no way to turn around."

I obviously survived. A few minutes later, back on planet earth, or what I'd call "level ground" a stream came into view. We had reached the other side. A stream flowed slowly along the valley floor, and children were laughing nearby and dangling their feet in the water.

Afterwards, we drove to the nearby Todra Gorge. It had a stunning formation of red rock walls rising vertically around us. Like the formation at Dades, millennia of tectonic forces had shaped the remarkable red cliffs that towered above us. Fortunately, there were no highways here trying to climb the lofty stone walls.

Like the Dades Gorge region, millions of years ago, this area was also submerged by the sea, when powerful tectonic forces pushed up the Atlas Mountain Range nearby. Today the river has its source in the highest point of the Atlas Mountains, and flows over 200 miles to the edge of the Sahara Desert.

Nearby, in the Dades Valley, there is a local village of Tamelait, which is world-famous for its roses. The harvesting of roses for perfume products is a significant event here. Each year there is a Rose Festival celebrated in this village. It is accompanied by a superb performance of the *Fantasia Horsemen* from Tinghir.

Young girl at the base of the Dades Gorge

Local villagers in the valley of Dades Gorge

World-Famous Twisting Highway in Atlas Mountains

View of Todra Gorge

Woman Engaged in Hand Crafts

Camels at the annual festival in the Dades Valley

Fantasia Horsemen at the Rose Festival in Dades Valley

Ancient Arch at Volubilis

Volubilis

One of the most interesting and unusual places in all of Morocco is the ancient Roman city of Volubilis, built in the third century BC. It was abandoned 1,400 years later in the eleventh century AD.

This city is located near two of Morocco's royal imperial cities, Meknes and Fes. In the beginning, Volubilis was part of the ancient civilization called Mauretania. At the height of its glory, it was an important trading center of the Roman Empire, with extraordinary architecture and impressive monuments.

Two other smaller sites of Roman ruins exist in other parts of Morocco at Sala Colonia and Lixus. Morocco was part of the Roman Empire until 500 AD.

Volubilis, however, was the most impressive of the Roman settlements. The city was built on a ridge overlooking the valley. Behind it, in the distance, was beautiful Zerhoun Mountain.

The ruler of Volubilis, Juba II, was appointed as king of Mauretania by Augustus. His impressive palace and royal court were here. Juba II was married to the daughter of Cleopatra and Mark Antony. Volubilis was one of the wealthiest of cities, with a population of about 20 thousand inhabitants.

At that time, five forts surrounded the city with several surveillance towers, to fend off opposing Berber tribes. Other Berbers, however, lived peaceably within the confines of the fortress. Also, a small enclave of Jewish residents lived inside the Roman settlement in the third century AD.

Ironically, it was not war or competing tribes that signaled the end of the city, but rather an earthquake in the fourth century. The elegant streets and the graceful arches of the remarkable city, were buried in the wreckage.

After the near total destruction, a Christian settlement was built here in the sixth century. It was followed by an Arab settlement in the eighth century. These new cities were built south and west of the old city, leaving the original Roman ruins still buried beneath the rubble. Around these new settlements, a wall separated newer developments from the Roman ruins.

The eighth century Arab settlement had been founded nearby by Moulay Idris ibn Abdallah, who was a descendant of the prophet Mohammed. He began what was later called the Idrisid Dynasty. He had arrived from Syria and became an Imam at Volubilis. Over a three-year period, he conquered most of Morocco in the north. He also built the city of Fes. He was beloved by his people, but his influence was cut short near Volubilis when he was assassinated.

A severe earthquake centered in Lisbon Portugal in 1755 caused severe damage to the site at Volubilis.

Today, while the site bears the marks of earthquakes and time, some of the early monuments have been restored. Excavations and restoration efforts were carried out primarily by the French government. Extensive work by French archaeologists to restore the city began under the direction of Hubert Lyautey in 1915.

That same year, other excavations were conducted by a team led by Marcel and Jane Dieulafoy. Part of some later French archaeological work employed thousands of German prisoners of World War I to help in the restoration.

Reconstruction efforts were interrupted during World War II, but they later resumed in 1955. By that time, Morocco was no longer under French rule. Some of the archeological finds from Volubilis are now exhibited in the Rabat Archaeological Museum.

Within the narrow confines of what was once the great city of Volubilis, silence now falls on the lofty arches and ancient mosaics embedded in the ancient stone floors.

All around it, life goes on. Arab and Christian settlements have called the surrounding area home. But in this abandoned Roman outpost, the elegant stone walls keep their secrets. Inside the ancient city, built upon the sands of time, no one has lived here for a thousand years.

Woman overlooking archaeological site

Stone Portal at Volubilis

Visitors to the Volubilis site

Birds nesting in the ruins of ancient Volubilis

Visitors look at mosaics embedded in ancient floor.

Stone cutter, at a local shop in Erfoud, examines an archaeological specimen embedded in rock.

Erfoud

Erfoud is located in the Sahara Desert in the eastern part of Morocco. It has a population of about 24,000 people. The elevation is 2,700 feet.

Erfoud is known as the Gate to the Sahara, and it is the staging point for camel forays into the dazzling red deserts of this region. Spectacular sunsets provide a view of reddish town walls, and gigantic swirling dunes rising from the desert floor.

The city is an oasis. It is divided into six districts or neighborhoods referred to as "Hays". These include Salam, Ziz, El Bathaa, Jdid, El Hamri and Annahda.

The city, itself, was built by the French a century ago. This small town has the only Royal Palace in Morocco that was built in the Sahara Desert itself. Erfoud is driving distance from the cities of Ouarzazate, Tineghar, and Fes.

Erfoud is located near the Erg Chebbi Dunes where visitors often search for ancient fossils.

Erfoud is known for its active trade in fossils. Workshops employ fossil experts to find and prepare fossils for sale. These range from small specimens embedded in stone to large full-size dinosaur skeletons. Reconstructing and polishing of fossils and specimens rely on the exacting and methodical skills of local craftsmen.

The Ksar Ksir Siffa is a Fossil Museum with local fossil finds on display.

Restaurants and tourist hotels are available in Erfoud, as well as in the neighboring desert town of Merzouga.

Erfoud has also provided a backdrop for a number of film projects. Producer Allan Cameron constructed a set near Erfoud to film "The Mummy". Also, British film makers used the stunning desert background for their movie production of the "Prince of Persia".

Camel treks are popular with visitors from the West. One-day or three-day treks with tent accommodations are possible options for tourists.

A local business called the Macro Fossils Kasbah is an ideal place to see authentic ancient fossils, and watch the artisans cleaning, polishing and cutting the fossils for museums or for local tourist shops. Trilobites embedded in stone, and other interesting finds, can easily be millions of years old.

Local restaurants offer popular Moroccan dishes, and often showcase native performers and traditional dances. Local cuisine can be found in the Restaurant Alt Ben Assou, and the Restaurant des Dunes.

Olives, fruit and dates are also available in small stands in the medina, and sometimes sold by local families. Popular specialties with street vendors include snails in soups or snails mixed with herbs. Also, an enduring favorite specialty here is Moroccan mint tea which is always popular any time of day.

Each year in late autumn, there is a Palm Date Festival in Erfoud. It is a three-day music event and very popular with local residents.

Workyard with several excavated rock specimens waiting to be identified

Tyrannosaurus Rex Dinosaur Skeleton

Village Shop in Erfoud

Tourist Tents

A Woman at the Well

Afternoon in the Desert

Mother and her children inside Desert Tent

Tourist Camps on the Sahara Desert

Young Child in Tent

Children in their tent at Erfoud

Man Standing on a Sand Dune

Printed in the United States
by Baker & Taylor Publisher Services